WHY DO PEOPLE BELIEVE IN GLOBAL WARMING?

A SIMPLE COUNTRY BOY EXPLAINS IN LAYMAN'S TERMS

INTRODUCTION

"....because "God's still up there", the "arrogance of people to think that we, human beings, would be able to change what He is doing in the climate is to me outrageous."[....]

Oklahoma Senator Jim Inhofe Chairman of Senate Environment Committee.

He also brought in a snowball on the Senate floor as a way to demonstrate that global warming must not be true since if it were there'd be no snow.

Nice reasoning there.

"The concept of global warming was created by and for the Chinese in order to make U.S. manufacturing non-competitive"

Donald Trump tweet 6 Nov 2012

This is one of many, many tweets the president elect has tweeted explaining how global warming is a hoax and an expensive one at that.

"The emissions that are being put in the air by that volcano are a thousand years' worth of emissions that would come from all of the vehicles, all of the manufacturing in Europe."

Senator Lisa Murkowski, (R-AK) – Chairman, Energy & Natural Resources Committee.

Her implication here being, that if volcanos put out more emissions than thousands of years' worth of vehicles, then how mankind could be contributing to warming of the earth.

"Calling CO2 a pollutant is doing a disservice the country, and I believe a disservice to the world."

Ex-Governor Rick Perry (R-TX).

Former governor of Texas, the nation's second largest state and once a presidential candidate. If CO2 isn't a pollutant like he says then how can it be bad?

As a side note: he should probably tell the US Navy this because each and every submarine has a CO2 scrubber that has as its entire job to remove CO2 from the subs internal atmosphere and there is a person on watch whose job it is to every hour walk by and take readings on this machine.

He should also tell Ron Howard, because he has an important scene in the movie "Apollo 13" where the astronauts try to create a CO_2 filter by using things they have onboard. (Like socks). Remember the scene where the guy throws random stuff on the table and says

"Make this, out of that, but using these"

One of the better scenes in that movie.

"I do not believe that human activity is causing these dramatic changes to our climate the way these scientists are portraying it."

Senator Marco Rubio (R-FL)

At least he is acknowledging that there are some dramatic changes going on in the climate. Few of his party even concede this fact.

"How long will it take for the sea level to rise two feet? I mean, think about it, if your ice cube melts in your glass it doesn't overflow; it's displacement. I mean, this is some of the things

they're talking about mathematically and scientifically don't make sense."

Ex-Rep. Steve Stockman (R-TX),

Wow! A complete misunderstanding of the difference between the Artic (which is floating ice) and Antarctica/Greenland (which is a land continent that has a lot of ice ON it) and the effects of melting ice on either and what the difference would be.

A lot of politicians seem to think a lot of climate scientist are full of crap when it comes to global warming.

In 2007, 498 member of American Meteorological Society and American Geophysical Union:

-97% of the scientists surveyed agreed that global temperatures had increased during the past 100 years

-84% said they personally believed human-induced warming was occurring

-74% agreed that "currently available scientific evidence" substantiated its occurrence.

This is the source of the often heard mantra that "97% of climate scientist believe in global warming".

Analyzed published research on global warming and climate change between 1991 and 2012 and found that of the 13,950 articles in peer-reviewed journals, only 24 rejected anthropogenic global warming. A 99.8% agreement in peer reviewed journals that global warming is a real phenomenon.

 A follow-up analysis looking at 2,258 peer-reviewed climate articles with 9,136 authors published between November 2012 and December 2013 revealed that only one of the 9,136 authors rejected anthropogenic global warming. A near unanimous agreement.

Who is right?

Politicians who often admit they are not scientist OR actual climate scientist?

The global warming opponents would say that climatologist are only spreading that message so they can get government funding for their work. In other words they are in it for the money?

BUT

Most global warming opponents are backed by oil and gas interest, since their product is the source of the cause of global warming so they have a vest financial interest to see that little or nothing is done in regards to global warming.

And who has more money

-government funded scientist

Or

-oil and gas interest

That choice seems easy. If money is supposedly the big driver behind this issue it's hard to believe that the scientist would "win" here.

But "Appeal to Authority" is no way to answer a question about objective reality anyway. Just because someone says they are a scientist doesn't mean they necessarily are saying something true.

What credentials does it take to be a climatologist anyway?

Climatology is defined as "the science of studying the climate

Well what defines that period of time? 100 years? 500?

Why is this introduction full of so many questions but no answers?

Over the course of the next several chapters I will try to present in layman-simple-as easy-as-I-can-understand terms what is the science behind global warming. Explaining why and on what grounds do very highly trained and educated people commit their lives work to study the climate, how do they do it and why have most of them concluded that YES the earth's mean (average) temperature is increasing, and YES it is increasing in a way that has not been seen during natural cycles over this period of time (150 years or since the Industrial Revolution), and YES mankind is one of the chief reasons behind this.

At the end of this at least you will know the scientist side of things (and NO it is not just temperature readings over the past 150 years although that is an important data source) and compare that to political rhetoric and see which side makes more sense.

You don't need to be a climatologist yourself or even very educated in the sciences to fully understand why they come to the conclusions that they do.

But you do have to take the time to read and listen to it.

In today's instant, iPhone Google ready world, it may take some intellectual time to consume the depth of the facts and science.

If you can commit to that you can get past the political noise and bias.

A good place to start to level set us would be to understand what "science" and the "scientific method" itself are exactly.

Contrary to popular belief science isn't just a litany or gathering of facts about nature that you have to rote memorize to pass Biology 101 in high school.

Science is a systemic way of observing, problem solving and thinking about nature and the universe we live in. We place our confidence in science by how well it can predict and explain nature in such a way that we can put it to use.

Climatology is one branch of science.

How science goes about this systemic approach to nature and how to problem solve is what we call the Scientific Method.

THE SCIENTIFIC METHOD

Right about now you're going to wish you paid more attention in middle school. (Don't we all.)

Many people think that Science is just a collection of various facts, terms etc. on certain subjects like physics, chemistry, biology, etc.

Additionally people think Science is what those geeky people do who can't attract sexual partners or carry a conversation at a party.

While Science is both of those things, it is also a body of knowledge and THINKING METHOD. Meaning how human beings go about solving problems in the real world and explaining how the universe works. That rigorous disciplined manner of thinking is called the Scientific Method and the knowledge it delivers we call Science.

In order to ground this discussion we should define some common terms and their use in Science because this is going to come in very handy as we go forward in this book.

I'm sure you've heard many people criticize things like evolution or global warming by saying "that's just a theory".

The implication being that since it's "just a theory" and not a firm hard never wrong "Law" then we shouldn't put too much stock into it.

Well something else that is "just a theory":

1. Germ theory of disease---the theory that microorganisms like bacteria and virus cause certain diseases. Is there anyone who doubts this today
2. Theory of electricity—the theory that electricity which lights our light bulbs and powers our world is the result of the flow of little bitty particles called electrons
3. Atomic theory—theory that all matter is made up of small things called atoms
4. Theory of Relativity—everyone has heard of this one

And on and on...

Why do some theories seem widely accepted while others are ridiculed?

Why aren't the widely accepted theories now called "laws"?

What exactly is a "law"?

Lack of understanding in these terms leads to a lot of misunderstanding in the general public on very important scientific concepts.

Let's start with these definitions

   A. Observation—actually seeing, hearing, touching, etc. a particular thing in the real world

   B. Fact—A type of observation. i.e. - fire is hot. It is Wednesday. Ten is greater than eight.

   C. Measurement—quantifying a fact. i.e. - that man is 200 lbs. The sun is 93M miles from earth.

   D. Data- a series of measurements and facts assembled together on a particular subject

Ok all those are easy enough. Now it gets a little trickier

   E. Hypothesis- an educated guess about WHY certain facts, observations, data are the way they are. This is the start of the way human beings go about explaining via a narrative or "story" on how our universe works. It uses facts/data and the things we observe to put it all together so to speak. A hypothesis is MEANT to be tested and it is tested by follow through with the logic of the guess to produce PREDICTIONS.

i.e. - that man weighs 200 lbs. because he eats more and exercises less today than we he did when he weighed 150 lbs.

So "eating more and exercising less" is our hypothesis on why he has gained weight from 150 to 200 lbs.

We can test this by follow up on the logic. If it is true, then we PREDICT that if he ate less and exercised more his weight then will go down.

Were we to carry out this experiment we can see if our hypothesis stands this test or not.

If it does not then we reject it and we create another hypothesis.

If it does we now have confidence in the hypothesis and seek to develop deeper analysis, more predictions etc.

(Note: this does NOT mean that we have "proven" the hypothesis. The reason is that it could possibly be another yet unstated or un-thought of reason that caused this effect to occur rather than our hypothesis. We have not ruled that out as of yet.)

If we can produce two or more hypothesis and have them stand up to rigorous tests then we can develop a

  F. Theory—a theory is a lot like a hypothesis in that it is a
     narrative as well, a good way to think about it is:

i.)   A hypothesis strings together many facts and observations in an explanation

ii.)   A theory strings together many hypothesis in an explanation

It's important to note at this point that a hypothesis and theory can both be disproved but neither can ever be PROVEN. That's right a theory, no theory ever, can be proven.

Why is that?

Because a hypothesis and theory are just our human explanations about how nature works but it could be that nature works a different way BUT produces the same result we see at our level of resolution and our hypothesis and theory just happens to explain this small section of the universe as well.

Picture it like this, in a trial a person can be found guilty or not guilty but they are NEVER found INNOCENT. Innocent and not guilty are not the same thing. Innocent means you didn't actually do it in reality while not guilty means that the state just wasn't able to prove that you did the crime.

(Note: it can be possible to sometimes prove someone innocent in some case such as the person was somewhere else at the time, etc., that need not be what the accused needs to do in order to be found not guilty.)

Scientist don't look to go out and prove hypothesis and theories true. They just test them again and again with greater and greater detail.

That's how the Scientific Method and Science works.

This brings us to a "law"

G. Laws- are just a description that captures, quantifies the interaction between observable facts. Laws almost always come in the form of equations. Newton's Second Law of Motion is F=MA.
It basically shows the relationship between the mass of an object and how fast it will move as you apply more and more force to it.
Ohm's Law is V=IR which shows there is a relationship between the voltage, current and resistance in an electrical circuit.

Notice Laws never explain WHY the universe behaves the way it does.

Consider E=MC^2. This is viewed as a law. It has been observed and measured many times over and over. It is the THEORY of Relativity that attempts to explain WHY it's true.

Newton's Law of Gravity is

$$F = (M1)*(M2)*G/r^2$$

This is an observable fact and can be replicated and repeated by anyone

Why it is the way it is, is addressed in Einstein's Theory of General Relativity. (Mass of a body curving space-time).

An important thing to keep in mind that no amount of evidence, experimentation can ever prove a theory OR turn it into a law. That's just now how it works.

The criticism then that "that's just a theory" is just vapid and meaningless and shows a lack of understanding of the Scientific Method by the person laying that claim.

If you are one of those people then please stop it now!!!

The Scientific Method then works like this:

1. We make observations of the world around us using our senses
2. We make measurements of these observations in a formal way
3. We develop a hypothesis which explains why these observations and measures are the way they are

4. We test this hypothesis by either conducting a controlled experiment (if we can) OR making prediction about the logical implications of the hypothesis
5. If the prediction turns out to be false then we reject our hypothesis and form another one
6. If the prediction turns out to be true we develop additional hypothesis that explain more of the observations and data and in a way that gives us deeper understanding.
7. If we have several hypothesis that survive these test then we form a more detailed theory

That's it.

Not really that complicated is it.

At least on paper. In real life it can be quite complicated.

We have discussed things like data, fact, theory, etc. A good way to keep up with how this works is via a picture below:

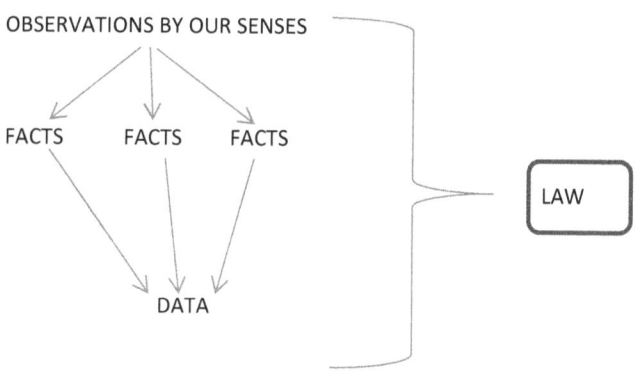

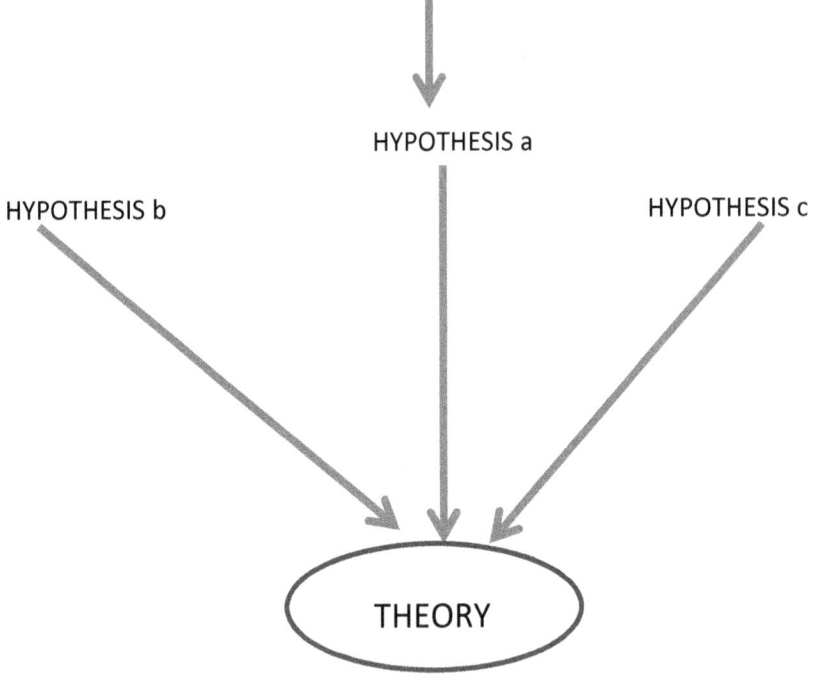

GREENHOUSE EFFECT

Much of Global Warming theory is based on something called "the greenhouse effect". That is basically what it sounds like.

Meaning that the earth is warmed much like a greenhouse.

So what is a greenhouse?

If you are not familiar it is a structure usually made of glass or some other type of transparent material (clear plastic for example), that lets the sun's rays which heats the inside and keeps plants warm, while at the same time is insulated such that it does not lose heat via convection (when one warmer body touches a cooler body and thereby transfers heat from hot to cold).

Picture your parked car in a mall parking lot in spring or autumn. While it may only be 75 deg. or so; not that hot, actually comfortable to most people; the inside of your car can get so hot as to kill any pets or children you may leave in there.

If the outside temperature is only 75 deg. how is this possible?

Because the sun's rays can enter into the car via the windows and heat up the air, YET the air itself cannot transfer it's heat back to the outside because while the window will let the sun's ultraviolet warming rays in, the glass will NOT allow (or not easily allow) the heat to transfer by touch to the outside air just outside the car.

Hence your car heats up to over 100 deg. F in some cases while the outside air is only 75 deg. or so.

The website below gives a good illustration of how this works:

http://www.apollosunguard.com/what-happens-to-parked-cars-left-in-the-sun/

10 Minutes – Car Interior: 109°F

20 Minutes – Car Interior: 119°F

30 Minutes – Car Interior: 124°F

1 Hour – Car Interior: 133°F

90 Minutes – Car Interior: 138°F

As you can see a relatively comfortable 80 deg. F day can in only 90 min. lead to a very dangerous 138 deg. F.

This is called the Greenhouse Effect as this is how greenhouses themselves actually work.

It is also how the earth itself works.

The earth's atmosphere acts like the car windows or a greenhouse's glass in the effect of letting in the sun's ultraviolet rays which heat the earth up. The earth's now warm body in the form of the ground, dirt, all matter then radiates via infrared radiation (we will talk more about this later), heat back into space; especially at night when the now warm earth/ground is facing open space and not the sun.

BUT, some of this irradiated heat is blocked by the earth's atmosphere and it's a good thing. This blocking of that irradiation of heat from the earth back into space is what keeps the earth at a temperature at which human, plants and animals can live. Without this the earth would be a cold dead body.

As the moon is precisely.

The moon is roughly the same distance to the sun as the earth is, yet the moon is a cold lifeless place. It is too small for its gravity to hold an atmosphere and without an atmosphere it can't keep any heat it receives from the sun thereby irradiates all of its warmth from its mass back out into space and thus is very cold. Too cold to support life. (Which it couldn't have any way with no atmosphere.)

Below is a good illustration of how this works precisely in the earth's atmosphere:

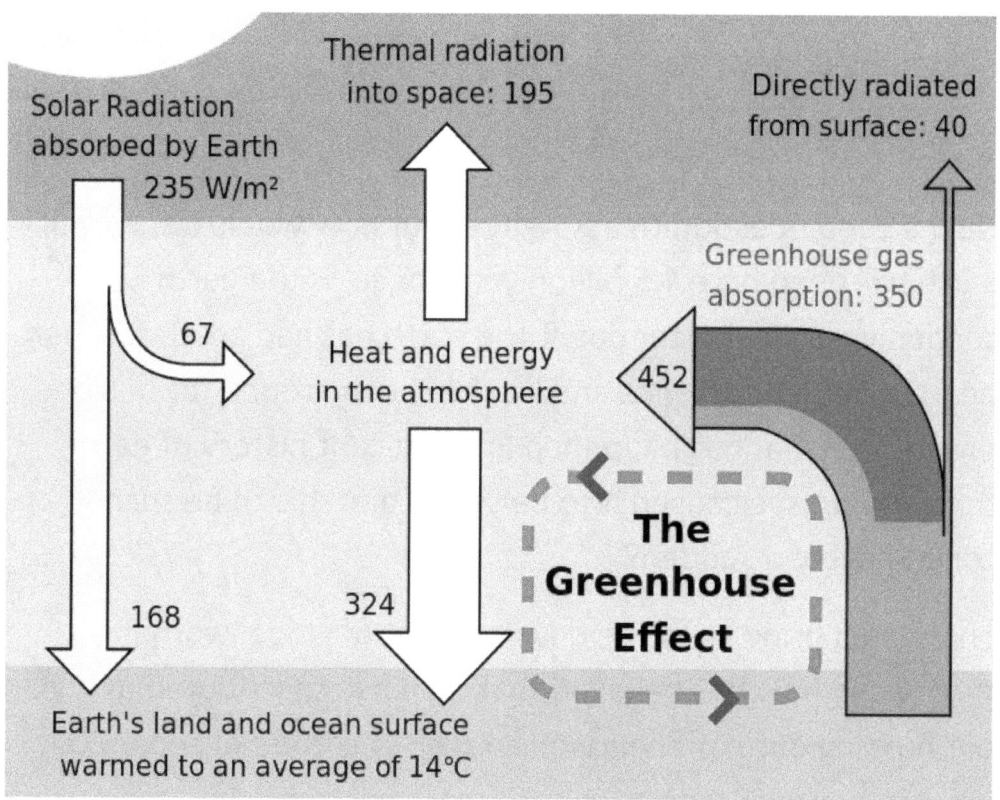

Following the algebraic math can be complicated but it is possible to account for all the "energy going into" the earth and the "energy going out" and thereby the difference would logically be the energy that would be heating the earth up.

Now a good question to ask right about now would be, why, if earth has been here 4.5 billion years or so, is the earth suddenly getting hotter now? The earth has had an atmosphere nearly all of its existence and live has been around for millions of year; so what has happened in the recent history of earth that would have caused it to be so hot and this to be such a concern all of a sudden?

This would bring us to the development of global warming theory and how it is believed that man has contributed to it via our burning of fossil fuels which releases carbon dioxide, CO2, into the atmosphere and it is this particular item in our atmosphere that is believed to be the main culprit of why we are seeing a warming earth.

HISTORY OF GLOBAL WARMING THEORY

The theory of global warming is by no means new.

As far back as the 1850's a British physicist named John Tyndall was able to discern the effects of the earth's atmosphere on the temperature of the earth. Tyndall explained the heat in the Earth's atmosphere in terms of the capacities of the various gases in the air to absorb radiant heat, also known as infrared radiation. Prior to Tyndall it was widely surmised that the Earth's atmosphere has a Greenhouse Effect, but he was the first to prove it. The proof was that water vapor strongly absorbed infrared radiation.

In the 1890's a famous Swedish physicist/chemist Svante Arrhenius known for his many contributions to the field of physical chemistry and thermodynamics and as a Nobel Prize winner in 1903, laid down the fundamentals on what we know today as global warming theory.

He explained how the greenhouse gas carbon dioxide, CO2 was related to air temperature.

While working on a theory to explain past ice ages he was the first scientist to attempt to calculate how changes in the levels of carbon dioxide in the atmosphere could alter the surface temperature through the greenhouse effect.

Arrhenius used the infrared observations of the moon to calculate the absorption of infrared radiation by atmospheric CO2 and water vapor.

He formulated his greenhouse law. In its original form, Arrhenius' greenhouse law reads as follows:

"If the quantity of carbonic acid [CO2] increases in geometric progression, the augmentation of the temperature will increase nearly in arithmetic progression."

This tells us what we know today in that as CO2 concentration increases in the atmosphere, so must the atmospheric temperature.

The following equivalent formulation of Arrhenius' greenhouse law is still used today:

$$\Delta F = \alpha \ln (C / C0)$$

Here

-C is carbon dioxide (CO2) concentration measured in parts per million by volume (ppm)

-C0 denotes a baseline or unperturbed concentration of CO2

And

- ΔF is the radiative forcing, measured in watts per square meter. Basically an energy or power measurement.

Arrhenius was the first person to predict that emissions of carbon dioxide from the burning of fossil fuels and other combustion processes were large enough to cause global warming.

And this was in 1896!

(We now know today that CO2 is a positive feedback look in our atmosphere. This means that as the amount of CO2 goes up so does the temperature and that as the temperature of the atmosphere goes up so does the amount of CO2 it can hold.)

In 1938, a British engineer named Guy Stewart Collander compiled temperature readings from the 1900 century and correlated these measurements with known CO2 atmospheric measurements made at the same time. He concluded that over the previous fifty years the global land temperatures had increased, and proposed that this increase could be explained as an effect of the increase in carbon dioxide. His papers throughout the 1940s and 50s slowly convinced some other scientists of the need to conduct an organized research program on CO2 concentrations in the atmosphere.

In the 1950's an American scientist named Charles David Keeling (this is the person Al Gore talks about in his documentary "An Inconvenient Truth") established a place at the base on Mauna Loa in Hawaii, two miles (3,000 m) above sea level.

This was an area that was considered to be far enough removed from human day to day activity to get some reliable and consistent measurements on the earth's atmosphere.

Dr. Keeling started collecting carbon dioxide samples at the base in 1958 and by 1960, he had established that there are strong seasonal variations in carbon dioxide levels with peak levels reached in the late northern hemisphere winter. A reduction in carbon dioxide followed during spring and early summer each year as plant growth increased in the land-rich northern hemisphere. In 1961, Keeling produced data showing that carbon dioxide levels were rising steadily in what became known as the "Keeling Curve".

Dr. Keeling's research shows that the atmospheric concentration of carbon dioxide has grown from 315ppm in 1958 to 380 ppm in 2005 with increases correlated to fossil fuel emissions. This is a 20.6% increase in just 47 years.

If we look back to the mid 1800's; CO_2 was measured in the earth's atmosphere to be about 290ppm. Today that figure is 400ppm which is a 38% increase and most of that can be contributed to mankind due to burning fossil fuels since the mid 1800's from the Industrial Revolution and invention of the internal combustion engine.

(Later we will discuss how, with satellites, we know it is CO_2 from man-made sources that is the biggest contributor of this.)

At the same time the earth's mean temperature was determined to be 56.7 deg. F. in the mid 1800's. Today it is 58.5 deg. F.

A near 2 deg. F in a little over 150 deg. Compared to historic rises in temperature seen long ago this is quite large and unseen before.

We know what the main greenhouse gases in earth's atmosphere are.

They are in order or impact:

1. Water vapor (including in clouds) H20: 10k-50K ppm contributing anywhere from 30-75% of greenhouse effect

2. CO2: 400ppm (today it is 400ppm or "part per million" particles by volume in the atmosphere.). This contributes 9-26% of greenhouse effect

3. Methane, CH4: 1.8ppm, contributing 4-9%

4. Ozone, O3: 2-8ppm, contributing 3-7%.

To better summarize, this table captures all the data above making it easier to see:

	COMPOUND	ABUNDANCE	CONTRIBUTION
1	Water vapor H2O	10-50K ppm	36-72%
2	Carbon dioxide CO2	400 ppm	9-26%
3	Methane CH4	1.8 ppm	4-9%
4	Ozone O3	2-8 ppm	3-7%

Depending on what water vapor is doing, H2O, carbon dioxide CO2 is contributing anywhere from 9% (not that much perhaps but can be significant) to as much as 26% to the Greenhouse Effect which is influencing earth's mean temperature.

It's clear that H20 and CO2 are the most important molecules in the atmosphere that effect the Greenhouse Effect and hence global warming.

But how do scientist know how warm the earth has been over the years? The mercury thermometer that we are all familiar with wasn't invented until 1714 by Daniel Fahrenheit (whom our temperature scale is named after. By the way the US is one of very few countries in the world that still use this scale. Most do the decimal 10 base system of Celsius) in Amsterdam. How then can scientist know the earth's temperature 1,000 or 50,000 years ago?

To understand that we will have to delve more deeply into H2O and it's interaction in the atmosphere

H2O AND its MANY FORMS IN THE ATMOSPHERE

Everyone knows, or should, that H2O means

- 2 parts of hydrogen, H atom
- 1 part of oxygen, O atom

And Hydrogen, H, is hydrogen because it has 1 proton in its nucleus.

Oxygen, O, is oxygen because it has 8 proton in its nucleus.

But both hydrogen and oxygen have isotopes. If you will recall from high school chemistry or physics class, an isotope is an atom that has the same number of protons but different number of neutrons in its nucleus.

In the case of hydrogen, 99.8% occurring in nature is H with 1 proton. However there is also deuterium which is H2 and that is an atom with 1 proton and 1 neutron.

As well there is also tritium, H3, which is 1 proton and 2 neutrons in the nucleus.

All of these are still hydrogen as they all have 1 proton but a certain small percentage of hydrogen in the atmosphere is deuterium and tritium.

Tritium is radioactive with a half-life of 12.3 years (meaning that if you have 10 lbs. of tritium in 12.3 years half of it would have radioactively decayed into something else or 5 lbs. Specifically Helium 3, He3. This radioactive decay when done with Carbon, C14, this is one way to date certain items.)

Tritium is an important element in that it is used to provide the fusion explosion of modern day nuclear weapons.

Hydrogen by the way is the only element that has separate names like deuterium and tritium for its isotopes.

Likewise oxygen, O, has isotopes. Three of them are stable and labeled as O16, O17, and O18.

The reason for this isotopic chemistry lesson is that the isotopes of H and O in H2O is one way scientist are able to determine what the temperature in the earth's atmosphere was like thousands of years ago.

This is done by analyzing ice core samples.

ICE CORE SAMPLING

An ice core is a core sample that is typically removed from an ice sheet, most commonly from the polar ice caps of Antarctica, Greenland or from high mountain glaciers elsewhere.

As the ice forms from the incremental buildup of annual layers of snow, lower layers are older than upper, and an ice core contains ice formed over a range of years.

The properties of the ice can then be used to reconstruct a climatic record over the age range of the core, normally through isotopic analysis. This enables the reconstruction of local temperature records

Ice cores contain an abundance of information about climate. Inclusions in the snow of each year remain in the ice, such as wind-blown dust, ash, pollen, bubbles of atmospheric gas and radioactive substances.

The variety of climatic proxies is greater than in any other natural recorder of climate, such as tree rings or sediment layers. These include (proxies for) temperature, ocean volume, precipitation, chemistry and gas composition of the lower atmosphere, volcanic eruptions, solar variability, sea-surface productivity, desert extent and forest fires.

The length of the record depends on the depth of the ice core and varies from a few years up to 800,000 years.

The time resolution (i.e. the shortest time period which can be accurately distinguished) depends on the amount of annual snowfall, and reduces with depth as the ice compacts under the weight of layers accumulating on top of it. Usually it takes approximately 50 years of pressure from the weight of ice to solidify and form the parts of the ice core that makes it easy for scientist to determine the elements that may have been in the snow and atmosphere way back when.

Ice sheets are formed from snow. Because an ice sheet survives summer, the temperature in that location usually does not warm much above freezing. In many locations in Antarctica the air temperature is always well below the freezing point of water.

Ice cores have been taken from many locations around the world. Major efforts have taken place on Greenland and Antarctica

Basically in the Antarctica, the snow doesn't really melt year over year. At least not fully. By sticking a metallic rod down into the ice, scientist are basically "going back in time" so to speak.

Each year in winter a new snow fall, falls in Antarctica. During the summer the sunshine maybe melt some of the snow forming ice which of course re-freezes. As more and more snow piles onto previous snow, the layer of snow gets higher and higher. As a result the pressure from the weight of new snow compresses and compacts the snow from last year, the year before, etc.

When you stick the metallic rod into the snow and pull it out you can see each year's snow falls much like you can see the years in tree rings.

Some of these metallic rods are over a mile long which translates to hundreds of thousands of years each of which scientist can see.

In the images below you can see scientist drilling to get ice core samples.

Below you can see the various periods of snow fall (representing period's years for the most part) in the sample

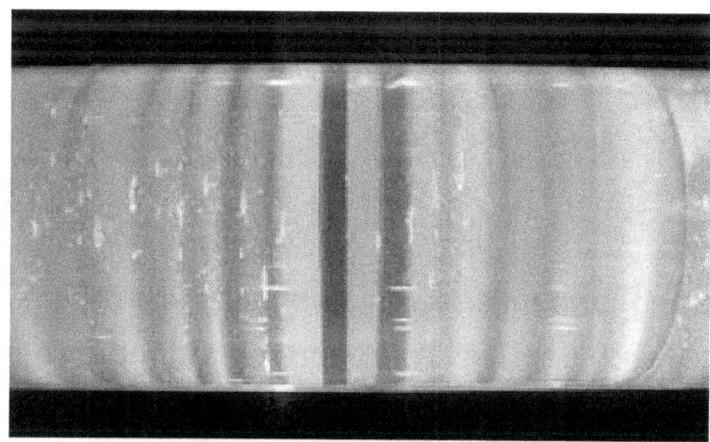

The properties of the ice can then be used to reconstruct a climatic record over the age range of the core, normally through isotopic analysis. This enables the reconstruction of local temperature records

The time resolution (i.e. the shortest time period which can be accurately distinguished) depends on the amount of annual snowfall, and reduces with depth as the ice compacts under the weight of layers accumulating on top of it (usually about 50 years).

Using these ice core samples scientist can look at how much H_2O and CO_2 were in the earth's atmosphere at the time of the snow fall.

Also they can tell how much of the H_2O has radioactive isotopes and by comparing the ratio of radioactive H_2O to non-radioactive H_2O climatologist can infer the temperature of the earth's atmosphere at the time.

The ratio (relative amount) of these two types of oxygen in water changes with the climate. By determining how the ratio of heavy and light oxygen in marine sediments, ice cores, or fossils is different from a universally accepted standard, scientists can learn something about climate changes that have occurred in the past

Evaporation and condensation are the two processes that most influence the ratio of heavy oxygen to light oxygen in the oceans. Water molecules are made up of two hydrogen atoms and one oxygen atom. Water molecules containing light oxygen evaporate slightly more readily than water molecules containing a heavy oxygen atom.

At the same time, water vapor molecules containing the heavy variety of oxygen condense more readily.

In polar ice cores, the measurement is relatively simple: less heavy oxygen in the frozen water means that temperatures were cooler.

The cornerstone of the success achieved by ice core scientists reconstructing climate change over many thousands of years is the ability to measure past changes in both atmospheric greenhouse gas concentrations and temperature. The measurement of the gas composition is direct: trapped in deep ice cores are tiny bubbles of ancient air, which we can extract and analyze using mass spectrometers. Temperature, in contrast, is not measured directly, but is instead inferred from the isotopic composition of the water molecules released by melting the ice cores.

At a range of sites in the Polar Regions scientists have measured a near linear relationship between

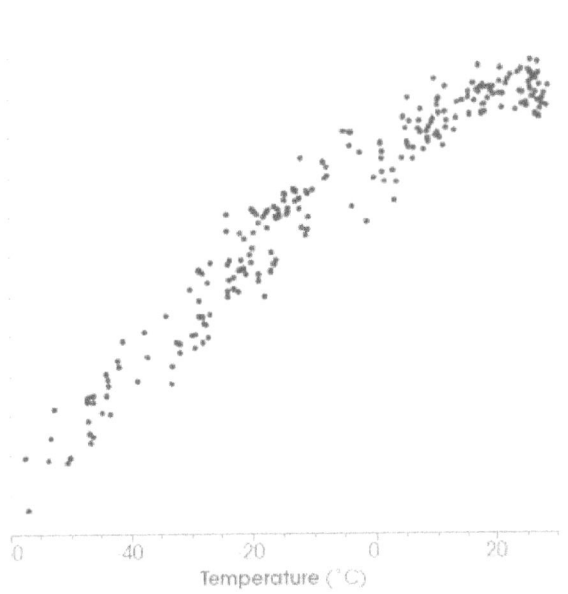

Temperature ($^\circ$C)

There is less O18 and Deuterium during cold periods than there is in warm. Why is this? Simply put, it takes more energy to evaporate the water molecules containing a heavy isotope from the surface of the ocean, and, as the moist air is transported pole-wards and cools, the water molecules containing heavier isotopes are preferentially lost in precipitation. Both of these processes, known as fractionation, are temperature dependent.

Bottom line: by analyzing ice core samples scientist can look at H_2O and its isotopes and can tell if one year is warmer than another year, so on and so on all the way up to present time when we can have actual temperature measurements of the earth's atmosphere and thereby construct a temperature vs time graph.

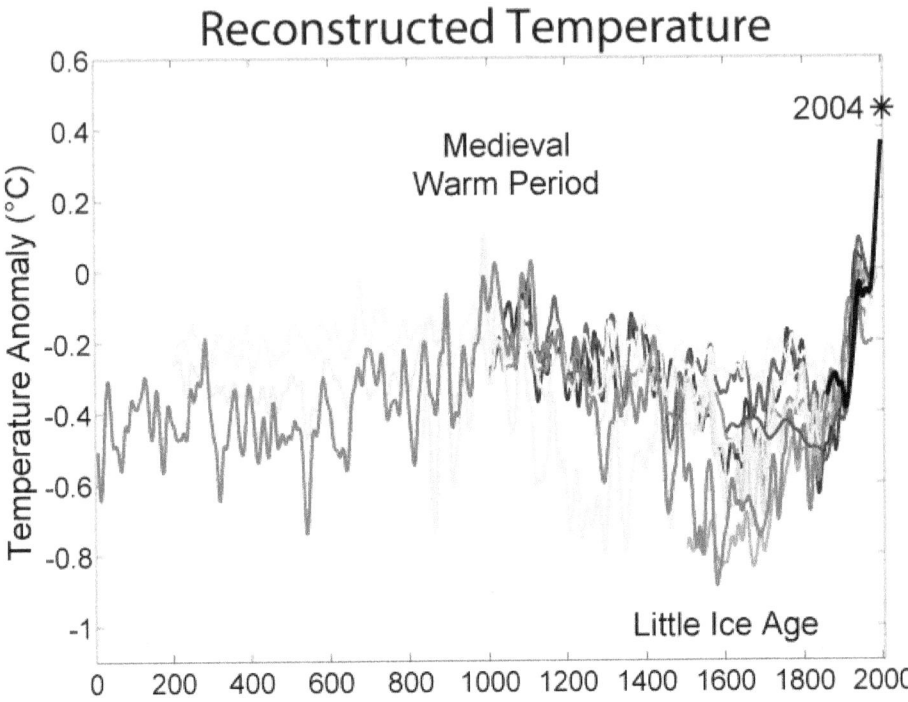

It is worth noting that the large spike you see from years 1800-2000 is larger than any other spike in any similar time frame over the past 2000 years.

Showing that at least over 2000 years this is NOT a normal temperature cyclical variation of the earth.

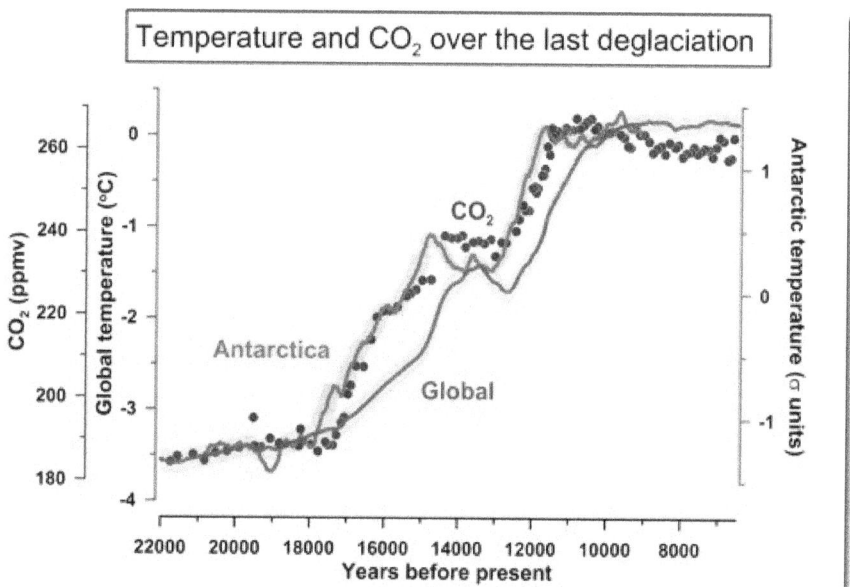

Temperature and CO_2 over the last deglaciation

Atmospheric CO_2 compared to Antarctic temperature and global mean temperature at the end of the last ice age. Note that global temperature is correlated with and generally lags behind CO_2, unlike Antarctic temperature, which slightly precedes CO_2. CO_2 data are from the EPICA Dome C ice core (Monnin et al., 2001, Science, 291, 112-114; Lemieux-Dudon et al., 2010, Quaternary Science Reviews, 29, 8-20), Antarctic temperature is based on a composite of five ice core records (Pedro et al., 2011, Climate of the Past, 7, 671-683), and global temperature represents an average of 80 proxy temperature records from around the world (Shakun et al., Nature, in press).

This type of graph is also consistent with the direct temperature by thermometer reading based on the last 100 years or so.

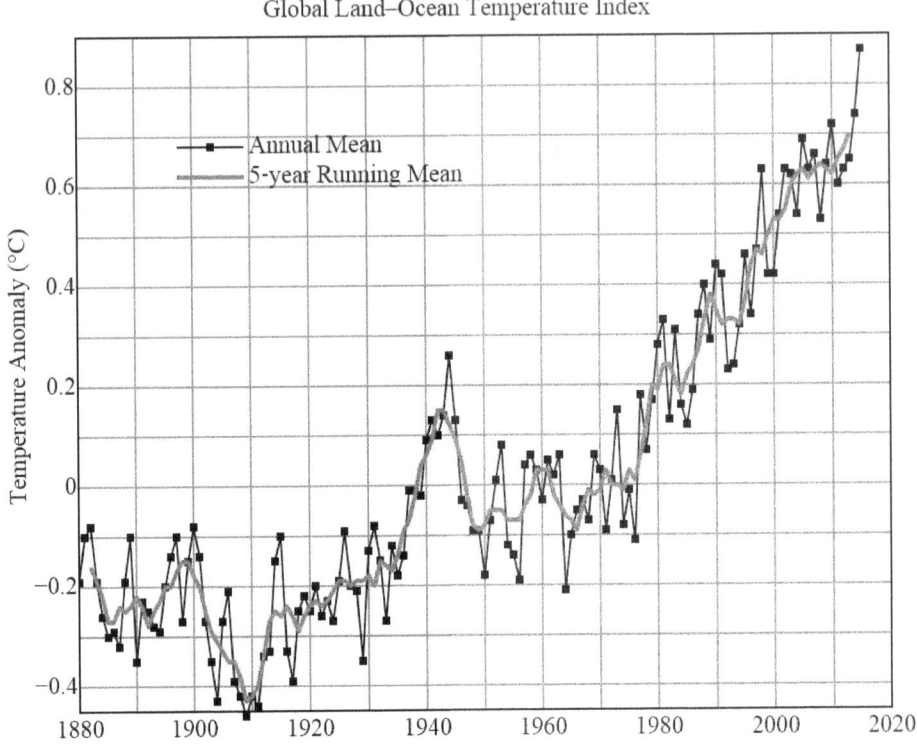

Global Land–Ocean Temperature Index

It is this type of data that allows climatologist to conclude that yes indeed the earth's mean temperature is going up, and if you look at the rate of temperature increase over such a relatively short amount of time (around 150 years) seems to imply that this is not a normal natural temperature variation.

Note that this argument is not just based on actual mercury temperature readings over the past 150 years but from data going back hundreds of thousands of years. The past 150 years just support and coincide with what all the other data says.

Another thing that allows scientists to know this the relationship between CO_2 over time and temperature. It is well known that humans have increased the amount of CO_2 in the atmosphere since the Industrial Revolution which happened to have started about 150 years or so ago.

CO2 AND its MANY FORMS IN THE ATMOSPHERE

Since 1800, CO_2 in the earth's atmosphere has risen 40% and because of the greenhouse effect, has warmed the planet.

The obvious source of the added carbon is the 330 billion tons of carbon that burning fossil fuels has added to the atmosphere since the Industrial Revolution.

Well then, let's prove it.

First, coal, oil, and natural gas come from plants and also have the distinctive carbon isotope ratio of plants. As CO2 in the atmosphere has built up steadily, its isotopic composition has shifted just as steadily in the direction of plant carbon. That tells us the added carbon is coming from plants. But what kind of plants? That question we can also answer.

One carbon isotope, C14, is radioactive and dies away to undetectable levels in 50,000 years or so. Fossil fuels, being millions of years old, have no C14 left. Adding ancient carbon should have lowered the proportion of C14 in the atmosphere—and it has.

For the last 50 years, as the amount of carbon in the atmosphere has increased, its C14 ratio has fallen steadily.

This is a tell-tale sign that it is human combustion that is putting this new CO2 into the atmosphere.

And it's not just ice core samples that allows us to see what CO_2 has done over the many years.

Another, quite independent way that we know that fossil fuel burning and land clearing specifically are responsible for the increase in CO_2 in the last 150 years is through the measurement of carbon isotopes.

If you will recall from last chapter isotopes are simply different atoms with the same chemical behavior (isotope means "same type") but with different masses. Carbon is composed of three different isotopes, C14, C13 and C12. C12 is the most common. C13 is about 1% of the total. C14 accounts for only about 1 in 1 trillion carbon atoms.

CO_2 produced from burning fossil fuels or burning forests has quite a different isotopic composition from CO_2 in the atmosphere. This is because plants have a preference for the lighter isotopes (C12 vs. C13); thus they have lower C13/C 12 ratios. Since fossil fuels are ultimately derived from ancient plants, plants and fossil fuels all have roughly the same C13/C12 ratio – about 2% lower than that of the atmosphere.

As CO_2 from these materials is released into, and mixes with, the atmosphere, the average C13/C12 ratio of the atmosphere decreases.

Isotope geochemists have developed time series of variations in the C14 and C13 concentrations of atmospheric CO_2. One of the methods used is to measure the C13/C12 in tree rings, and use this to infer those same ratios in atmospheric CO_2. This works because during photosynthesis, trees take up carbon from the atmosphere and lay this carbon down as plant organic material in the form of rings, providing a snapshot of the atmospheric composition of that time.

If the ratio of C13/C12 in atmospheric CO_2 goes up or down, so does the C13/C12 of the tree rings. This isn't to say that the tree rings have the same isotopic composition as the atmosphere – as noted above, plants have a preference for the lighter isotopes, but as long as that preference doesn't change much, the tree-ring changes will track the atmospheric changes.

Sequences of annual tree rings going back thousands of years have now been analyzed for their C13/C12 ratios. Because the age of each ring is precisely known we can make a graph of the atmospheric C13/C12 ratio vs. time.

What is found is at no time in the last 10,000 years are the C13/C12 ratios in the atmosphere as low as they are today. Furthermore, the C13/C12 ratios begin to decline dramatically just as the CO2 starts to increase — around 1850 AD or the time of the start roughly of the Industrial Revolution when humans really started burning fossil fuels for production of goods.

This is exactly what we expect if the increased CO2 is in fact due to fossil fuel burning.

Furthermore, we can trace the absorption of CO2 into the ocean by measuring the C13/C12 ratio of surface ocean waters.

While the data are not as complete as the tree ring data (we have only been making these measurements for a few decades) we observe what is expected: the surface ocean C13/C12 is decreasing.

Measurements of C13/12C on corals and sponges — whose carbonate shells reflect the ocean chemistry just as tree rings record the atmospheric chemistry — show that this decline began about the same time as in the atmosphere; that is, when human CO2 production began to accelerate in earnest.

In addition to the data from tree rings, there are also of measurements of the C13/C12 ratio in the CO2 trapped in ice cores. Just as we discussed in the last chapter on H2O measured in ice core samples from Antarctica and Greenland we can do the same with CO2.

The tree ring and ice core data both show that the total change in the C13/C12 ratio of the atmosphere since 1850 is about 0.15%. This sounds very small but is actually very large relative to natural variability.

 The results show that the full glacial-to-interglacial change in C13/C12 of the atmosphere — which took many thousand years — was about 0.03%, or about 5 times less than that observed in the last 150 years.

The lower ratio of radioactive CO2 in the atmosphere which also corresponds to the known amount of CO2 man has put into the atmosphere since the Industrial Revolution, is very compelling evidence that yes indeed it is human fossil fuel burning that is increasing the CO2 in the atmosphere and hence increase the earths mean temperature over this same time period.

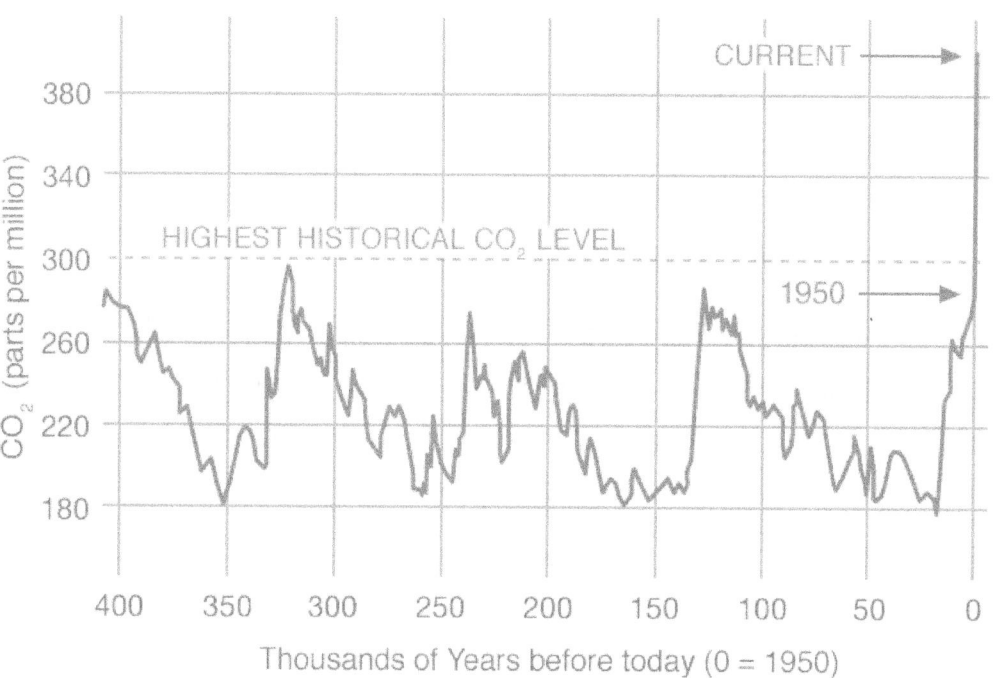

So we know CO2 effects temperature going all the way back to Arrhenius, we know that CO2 has been going up dramatically (see graph above), and we know that the CO2 increase is due to human activity by looking at the ratio of C13/C12 since this is what is exactly expected by burning of fossil fuels—ergo: humans are causing global warming.

One of the worse things about CO2 in the atmosphere is that it sticks around a long time.

There are many heat-trapping gases (from methane to water vapor), but CO2 puts us at the greatest risk of irreversible changes if it continues to accumulate unabated in the atmosphere. There are two key reasons why:

1. CO2 has a high "radiative forcing" (RF) compared to of each climate driver—in other words, the net increase (or decrease) in the amount of energy reaching Earth's surface attributable to that climate driver.

   Positive RF values represent average surface warming and negative values represent average surface cooling. CO2 has the highest positive RF (see table below) of all the human-influenced climate drivers compared by the scientist.

   Other gases have more potent heat-trapping ability molecule per molecule than CO2 (e.g. methane), but are simply far less abundant in the atmosphere and being added more slowly

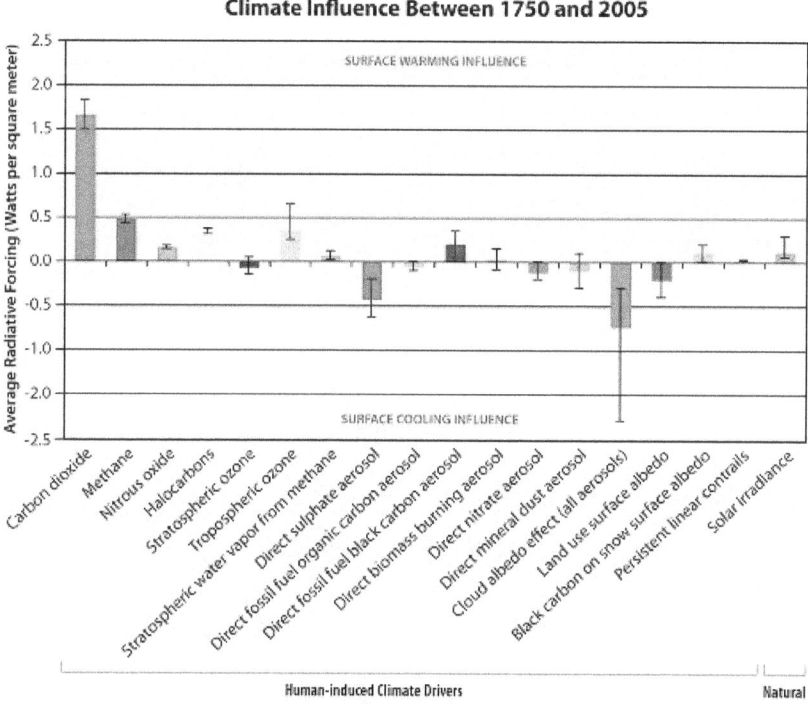

Climate Influence Between 1750 and 2005

Source: IPCC 2007 WGI Table 2.12; Figure: Union of Concerned Scientists

2. CO2 remains in the atmosphere longer than the other major heat-trapping gases emitted as a result of human activities. It takes about a decade for methane (CH4) emissions to leave the atmosphere (it converts into CO2) and about a century for nitrous oxide (N2O).

In the case of CO2, much of today's emissions will be gone in a century, but about 20 percent will still exist in the atmosphere approximately 800 years from now. This literally means that the heat-trapping emissions we release today from our cars and power plants are setting the climate our children and grandchildren will inherit. CO_2's long life in the atmosphere provides the clearest possible rationale for reducing our CO_2 emissions without delay

BUT earlier didn't we say that H_2O was by far the most abundant Greenhouse Gas in the atmosphere? Why isn't it such a concern when talking about Global Warming?

The principal reason is that water vapor has a short cycle in the atmosphere (a few days) before it is incorporated into weather events and falls to Earth, so it cannot build up in the atmosphere in the same way as carbon dioxide does. H_2O then comes and goes and changes with a great deal of volatility and involved with so many natural weather events, that it simply isn't a major concern for long term earth temperature.

It literally comes and goes and changes too much very fast.

Completely opposite of CO_2 in the atmosphere.

There is yet another way we know that CO_2 is the major variable effecting Global Warming.

Take a look at the picture below:

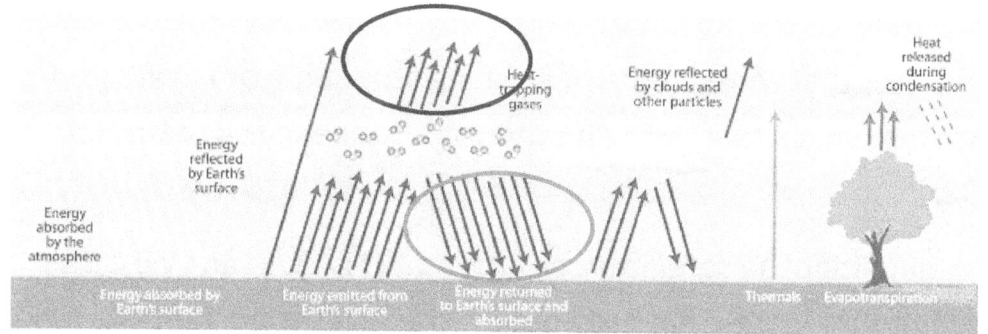

The top blue circle represents heat being irradiated back into space from the earth's surface after it has been heated by the sun.

This infrared heat has a thermal signature that can be captured by satellites and analyzed.

The bottom green circle represents that portion of heat radiated by the earth's surface that is stopped by heat trapping Greenhouse Gases like CO2, methane, etc.

When analyzing the irradiated heat being sent back into space, satellites can tell from the frequencies that are missing, just what molecule is stopping the heat from going into the atmosphere.

Every molecule has a resonance frequency. Infrared heat that matches that frequency will be stopped by said molecule in the atmosphere.

It is a fact that most of the heat being stopped from irradiated back into space matchers the thermal-frequency signature of CO_2.

Meaning that the satellites confirm that it is CO_2, and no other Greenhouse Gases that are keeping the biggest portion of earth's irradiated heat from going into space- and hence warming the earth up.

We know then that it is not sunspots, not cosmic space radiation, not volcanoes, not earth's precision, not a normal ice age cycle BUT CO_2 that is causing the warming of the earth over the past 150 years or so.

Furthermore we know that it is humans that have put that additional CO_2 into the atmosphere and that is causing this warming.

(By the way it is worth noting that the Greenhouse Gas, methane, is also increasing due to human activity. Two big sources of methane is from cattle waste-feces, flatulence, etc., as well as the biological activity of termites. Both of these have increased over time due to human activity such as domestication, growing of cows for our food supply, and termite activity due to deforestation actions, building of houses, and other wood turn over activity. So the two of the top three Greenhouse Gases increase can be laid at mankind's door step.)

SUMMARIZING ALL THIS

We've covered a lot of material some of it quite technical to laypeople (and here I said by the sub-title of my book I wouldn't do that) so let me now summarize everything in a pithy type of manner

-Climatologist is a legitimate science practiced by people with PhDs who are very dedicated. So dedicated that they live in harsh climates like Antarctica sometimes for many years

-Climatologist can tell what the earth's mean temperature has done going back hundreds of thousands of years using ice core samples

-This data shows that the earth is warming.

-not only is the earth warming it is doing so at a rate not seen in the past few thousand years

-CO2 is an important Greenhouse gas.

-Human kind has been pumping CO_2 into the atmosphere at a rate that is measurable since the Industrial Revolution starting circa 1850

-We know that CO_2 is correlated with atmospheric temperature by Arrhenius law as well as thousands of years of data

-We know that CO_2 is the biggest driver of the earth's warming due to satellite data from earth's irradiation

-Based on the ratio of CO_2 isotope change over time we know how much of this CO_2 is due to man and how much is natural

-Therefore we can tell man's contribution to CO_2 and hence overall global warming

See, that wasn't so bad was it?

Whether you full accept or believe in global warming or not, certainly by now you realize that the criticism that global warming is a purely political subject and not based on any science- that is objectively not true. In fact it is offensively false.

But there are a lot of criticisms from global warming. Where does that come from?

In the next section I will attempt to go through and answer some of the better known laments against global warming.

What you will find it is that the OPPONENTS of global warming are politically driven.

That they have no real hard dedicated science on their side

It is they who are driven and funded by money interest that stand to lose if global warming is real and understood to be a problem for humanity.

ANSWERING COMMON GLOBAL WARMING CRITICISMS

Depending on what poll and when it was taken, anywhere from 30-50% of Americans don't believe global warming is a real phenomenon.

If you break it down by party or religious affiliation those numbers go even frighteningly higher.

In the recent presidential debate and in fact during the entire campaign, the subject was rarely brought up.

What are some of the reasons cited for not believing in global warming?

Let's look at a few.

1. *"The earth is too big for mankind to destroy. We can't be changing the weather"*

This is usually trotted about by the very religious. The view that mankind could be so arrogant that he could destroy God's creation of the earth is practically heresy in their eyes.

Ironically they are correct. Mankind probably cannot destroy the entire earth. At least if by "destroy" you mean make the earth into a thousands of pieces of smoldering little boulders.

Humans do not have that capacity even if we detonated all our nuclear weapons in concert at once and leveraged all our technology.

But so what?

Global warming doesn't say otherwise. Global warming doesn't say that an increase in the earths mean temperature will destroy the earth. In fact in the earth's history it has been many times hotter than it is in modern day. CO2 has been 10-100x what it is in modern day.

Yet the earth still exist.

To be fair those conditions were hundreds of millions if not billions of years ago.

But the issue with Global warming is that human civilization has only been around a few thousand years. Most of our cities have existed for less than 2,000 years or so (and in the case of the US way less than that) and they were built at locations based on how people found: river levels and behavior, coastal lands, agricultural conditions, rainfall, etc.

Well if all that changes since the time when those cities were built doesn't it stand to reason that maybe those cities will suffer?

Places like New Orleans, Charleston, SC; Manhattan, Costa Rica, Bangladesh, Miami, many parts of Africa, etc.- are all located in areas that can be easily damaged beyond safe habitation if the sea levels rose and in some cases rose not by much.

All those people will be displaced and refugees (at best) in some cases. They will need to be repatriated.

The earth on whole however will survive on. In fact the earth as an inanimate object doesn't care at all about Global warming.

It's people, plants and animals that have to be concerned.

Of course long term people, plants and animals will adjust.

That old joke "we will grow oranges in Alaska" is true.

However in the interim there will be a lot of pain and suffering for millions of people in some heavily populated areas. That will be an ugly transition.

So yes it's true humans can't destroy the earth, but we don't need to make life very miserable in the immediate future for a lot of people

That's the concern of Global warming.

2. *"Climate changes all the time. We are just in a normal cycle and humans have nothing to do with it"*

It is true that the earth's climate has changed over the years. We have had ice ages, warming periods coming out of said ice ages; mini-ice ages, etc. etc.

But few period if any have changes this much in this short of a time period.

Take a look at some graphs below:

Shakun et al
Marcott et al
A1B
HadCRUT4
2100
Temperature Anomaly [°C]
Year (BC / AD)

A
B
Years (BP)
Temperature Anomaly (°C) (1961-1990 CE)

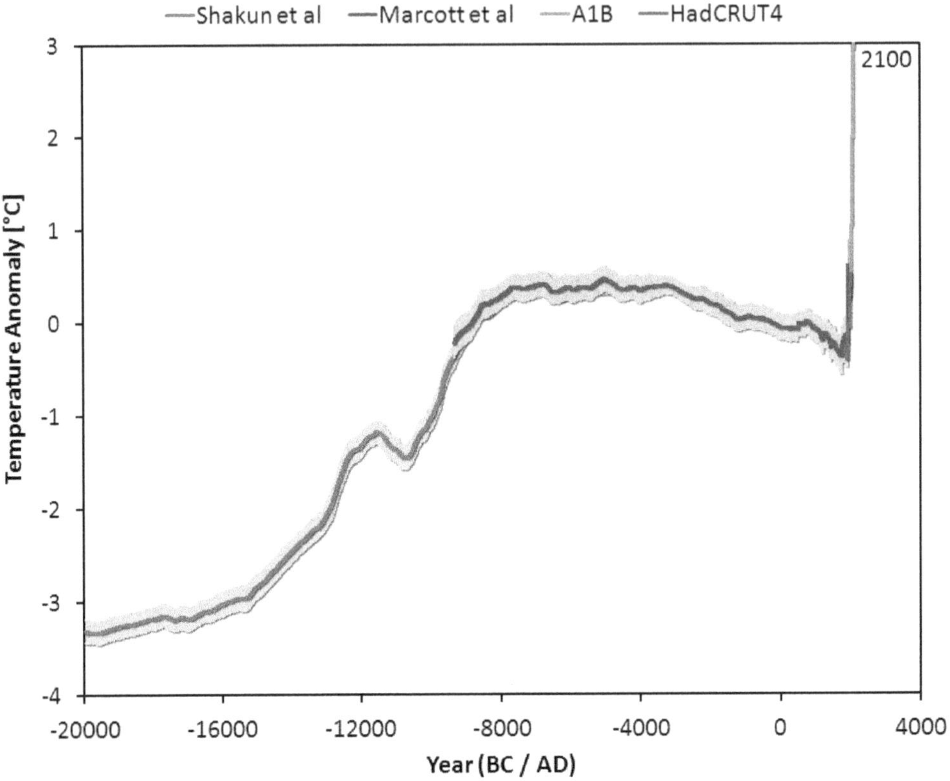

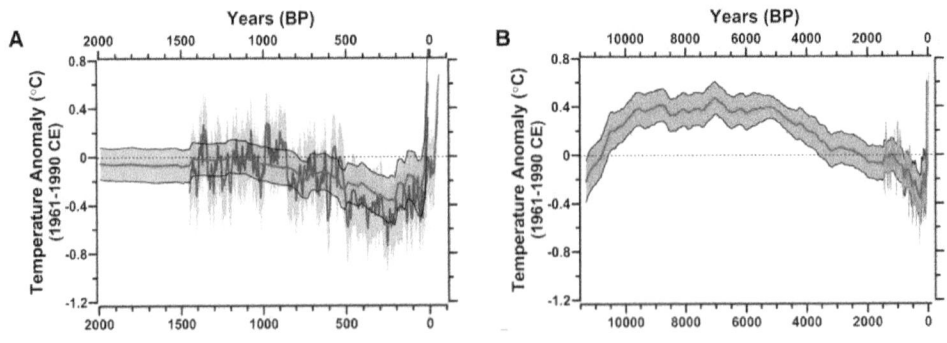

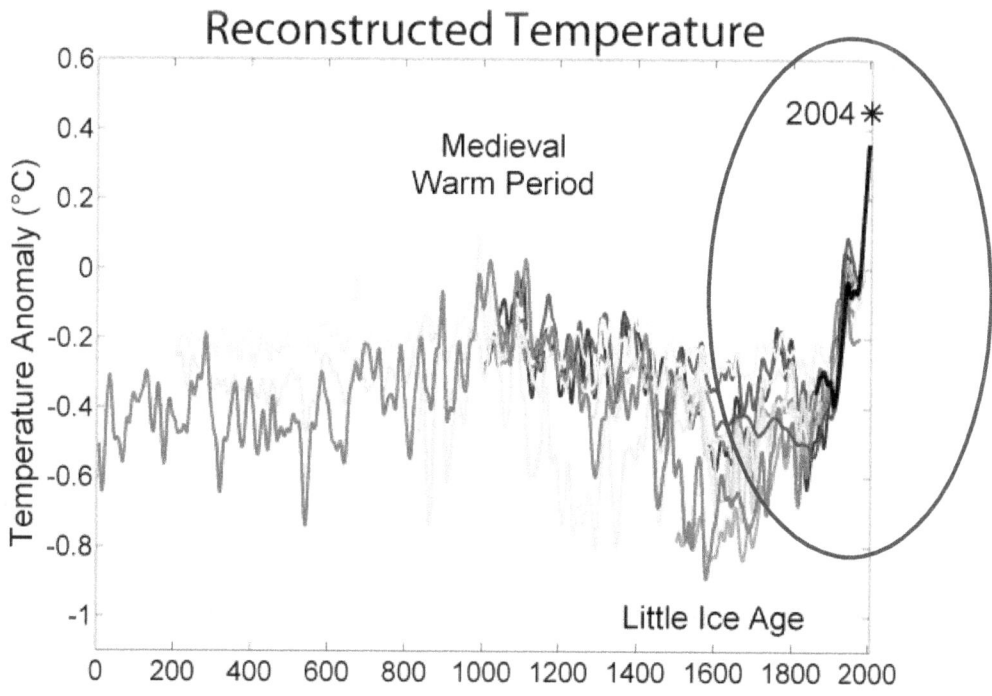

Notice the circle on the far right of the graph.

That shows that in less than 200 years (about 150 as we have discussed throughout this book) the temperature has risen by over 1 deg C.

If you searched this graph that goes back 2,000 years, you cannot find another 150 yr. period of time that has as high of temperature increase and this data has a Little Ice Age embedded in it.

This shows that over the past 2,000 years the natural variation in climate has not had a warming period equal to what has been produced since the Industrial Revolution.

NO this is not just normal climate variation.

3. *"CO2 only makes up 400ppm as in parts-per-million. It is so small in our atmosphere it can't possible have that big of an influence"*

It is true CO2 is 400ppm (which by the way is the highest it has been in over 6,000 years and maybe as much as 20,000 years) but without that so called small amount there would be no life on earth.

Not just because all plant life would die since plants inhale CO2 in order to live and if there were no plant life there would be no animal life soon afterward; but also because the planet would be too cold to support life as we know it.

People are often citing this under the guise that something in such a small amount could have such a big impact.

Consider however that .08% of alcohol in your blood makes one legally drunk. A very small amount when considering your entire blood volume.

Also a 160 lb. man if he ingested just .55 grams of cyanide will almost certainly die. That is just 76ppm proportionally much less than the CO_2 level in the atmosphere.

The point is there are many examples in nature where a small amount of certain substances or molecules can make a very large impact on a very large system.

4. *"The jury is still out in the scientific community regarding agreement on global warming"*

This is simply not true. Many people use surveys that usually include scientist outside the field of climatology and asking them open ended questions regarding global warming.

A simple survey is not the best way to look at this however.

To get a good idea of how the climate scientific community regard global warming look at peer reviewed published papers.

As I have stated previously almost 97-99% of all peer review articles in the past 15-20 years have been in support of global warming theory.

If you stick to the specialist in the field, climatologist, not necessarily chemist, biologist, physicist (although most of them actually believe in global warming as well it turns out) sociologist – just pure climatologist you get near unanimous consensus.

5. *"Scientist get funding from governments and the whole global warming is a hoax to get more funding"*

The "theory" (using that word very loosely) is that global warming is part of a political effort to raise taxes in a backdoor way or for socialist to take over via nationalization of industries by claiming said industry is harming us all via global warming therefore the government gives money for scientist to spread global warming via propaganda.

Frankly this is in line with conspiracy theorist such as those that believe 9/11 was an inside job or Sandy Hook was faked with child actors as an effort to bring on more gun control.

Several things make this "theory" preposterous.

For one thing scientist spend a lot of time measured in months and years in some pretty severe places such as Antarctica. In fact they are usually paid very little in proportion to their education. Most usually got PhD's and make well less than six figures.

If they are being paid for said message they are being UNDER paid.

However look at the people who advance this theory. Usually it is people associated who themselves are paid by the oil industry.

Why does that matter?

Because global warming theory points the finger at CO2 emissions which predominately comes from the burning of fossil fuels. That is oil.

So oil indeed has a vested interest in global warming being false because if it is true then there would be a big public effort via government or otherwise to do something about global warming and one of those things no double would probably involve reducing the burning of fossil fuels.

Meaning less oil sales.

And on which side do you think there is more money?

How often do you see scientist driving around in new convertible sports cars on the way to their big beach houses?

Rarely if ever.

But how many times have you seen oil executives and marketing VP's doing just that?

And how often do you see so called "scientist" spokesmen for the oil industry who are out in front of the global warming criticism, also doing that?

If you are one to be inclined to believe that money interest is driving this issue, then you certainly haven't got any logical reason to come down on the side of the "paid scientist" angle.

Furthermore and more importantly the bases of global warming made by scientist is REPEATABLE and REPRODUCIBLE.

This means that anyone can go to where this ice cores are stored and not only watch the scientist do their measurements and calculations but can even do it themselves if they have the ability.

The evidence is totally 100% transparent and everything is above board and can be verified by third parties in double blind studies as often is the case.

This criticisms is quite frankly silly and merits little response. Certainly a lot less than I have just given it.

6. "Global isn't that big a deal. We will just grow oranges in
   Alaska"

Interesting enough, though flippant, this actually has a merit of
truth to it.

In the long run mankind will no doubt adjust. We will indeed
grow oranges in Alaska and move as needed to different places.

But the interim will be very painful.

Between where we are now, where we got to go through
before we get to the new stable condition there will be a lot of
misery among our fellow humans.

People in low lying areas will have to move, perhaps millions of
people will be displaced and maybe become refugees. This will
put pressures on governments some of which already don't get
along with their neighbors, all making wars and conflicts more
likely.

Businesses like insurance, tourism will be harmed and many go
out of business leaving no telling how many people
unemployed and investments money lost.

While it's true that human beings will figure out a new normal
and we will come to live like that; it's just the transition period

we got to go through we will have to watch a lot of people suffer and die as the change takes place.

That is why so many people who support global warming want something done now. To ease this transition period (assuming that we can't stop what's already been done in regards to global warming) and make the suffering, misery and death less; it makes sense to take action earlier. Most problems people face are usually more easily and cheaply solved the quicker you find the problem and take action to mitigate it.

7. *"Global warming is false. There hasn't been any real warming in over 17 years"*

This is just a matter of purely cherry picking data. How critics do this is that they pick out an especially abnormal hot year 1998 and use that as the starting point and then compare subsequent years to show that there hasn't been any warming (there actually has even if you do this) or the warming has been too little to worry about.

There are two main problems with this criticism.

The first one is that this is only 17 years. When climate scientist talk about global warming they talk in terms of thousands of years.

Seventeen is too small of a period to draw any type of meaningful conclusion.

Climatologist have data going back 800, 000 years. And when you compare the last 150 years to all that data from the past 800,000 years you can clearly see warming going on since the mid 1800's.

The second issue is that this is a case of the Fallacy of "down the up stair case".

What this means is that global warming has been trending upward since the mid 1800's. This increase however isn't every single year, year after year.

It is a gradual trend based on 5 year averages, etc.

Few curves in life go up every single data point.

So what the critics do is pick an abnormally high year like 1998 when the temperature jumped a lot even more than the trend, and use that now as an anchor point so that when 1999, 2000, 2001 come in lower (but still on the gradual upward trend) it looks like that since 1998 temperatures have been going down.

Take a look at the graph below

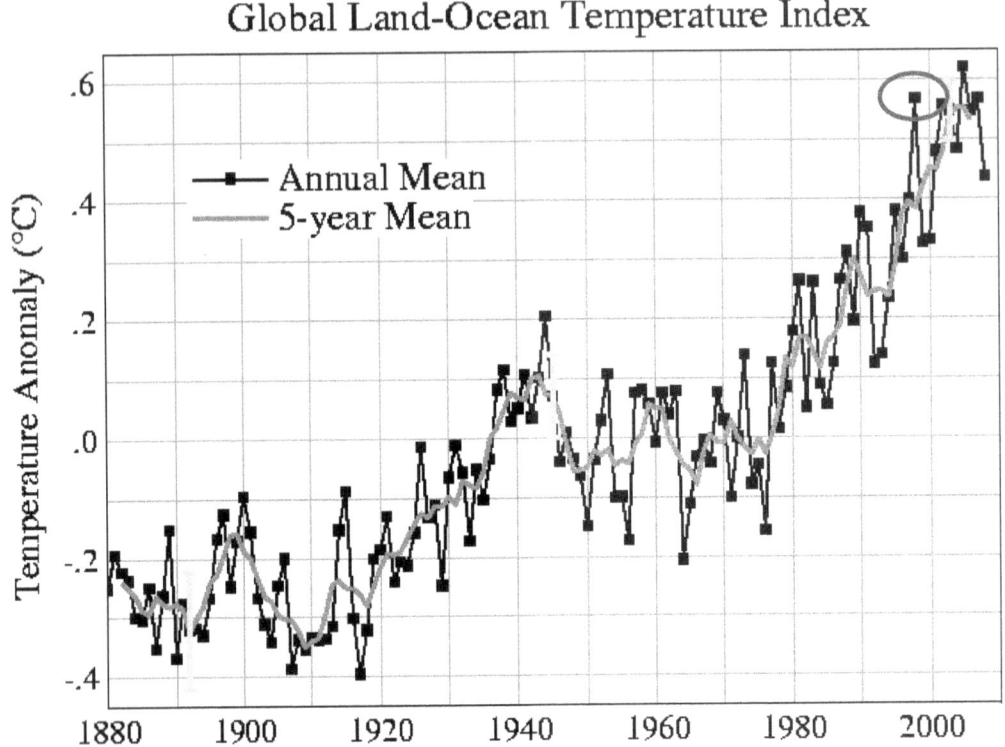

Global Land-Ocean Temperature Index

Clearly since roughly 1880-1900 this temperature trend has been going upward. Of course individual years jump around as there is still a lot of variation in the data. (More on that later.)

Take note of the circled area in the top right corner of this graph. That is 1998. Notice that it is exceptionally higher than the nearby years.

Well if one just were to look at 1998 and subsequent years, one could contrive a graph that implied that there has been no warming since 1998, or more convincingly the past 17 years.

However as you can see the upward trend continued even after 1998 and many years afterward were even hotter than 1998 but since the time frame is so low and resolution isn't sufficient to show up on a graph like this.

The black line is the temperature in individual years.

The red line is a 5 year rolling average.

When dealing with statistical data like this there is a saying that "the average of averages" gives you a better idea of how data is performing.

That is basically a fancy way of saying that keep your eye on the red line as it gives you a better long term variation and trend that is really going on.

As you can see the red line is still trending up even in the last 17 years as it has for the past 120 or so.

8. *"When there is a heat wave the global warming people blame global warming. Whenever there is a big snow storm they also black global warming. Which is it?"*

On its face this seems like a contradiction but actually it isn't.

Adding CO_2 into the atmosphere not only "shifts the mean" (makes the average temperature higher) it also increases the variation.

This is not just true with the earth's atmosphere but with any stable system. When you add energy into an otherwise stable system- and it doesn't matter what kind of energy be it thermal, gravitational, electromagnetic, etc. - you not only change the mean (average) but also the variation or standard deviation.

Consider the graph below.

One graph is a histogram/normal curve of say the earth's temperature for the past 2,000 years (this is just example data to make this point. Not actual temperatures). Called "Old Climate". This is meant to show the earth's temperature before humans began pumping out so much CO_2.

The other graph is earth temperature data since 1850. This is called "New Climate". It represents an earth's new mean temperature of 56 deg., two degrees higher than previous "Old Climate".

Also the variation has been increased. Before in the "Old Climate" the variation (actually its standard deviation which is the square root of the variation) was +-1 deg. The "New

Climate" due to the higher energy has a variation/standard deviation of +-5 deg. (4.6 deg exactly in this example.

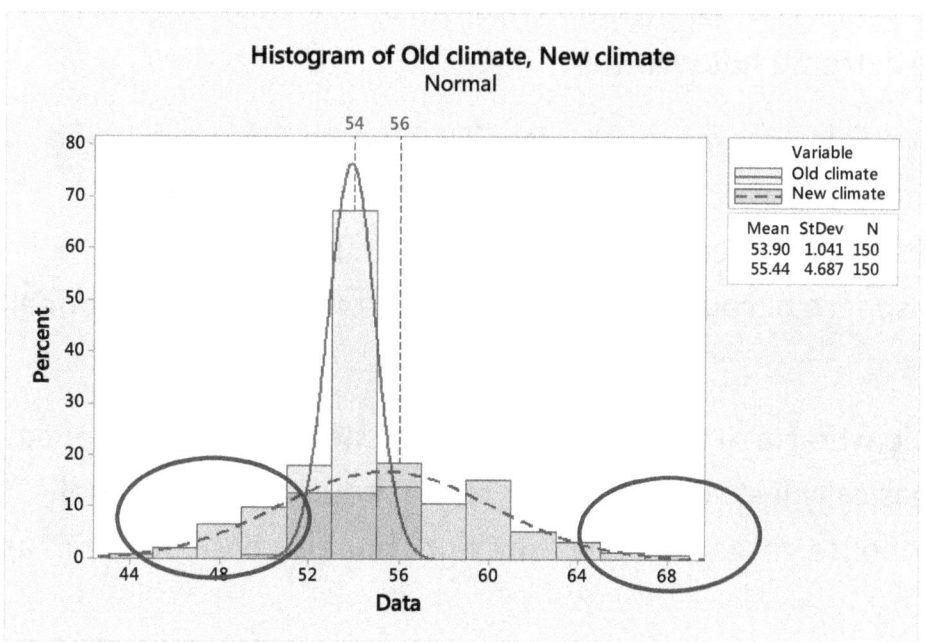

I bring up this example to make this point:

Notice the circled areas on the graph at the left and right tails.

These represent temperature extremes. On the right is really hot days. On the left is really cold days which would imply more and bigger snow storms.

The "New Climate" not only has gotten a higher mean; i.e. - global warming" but also has gotten a higher variation and as a result of that higher variation it has ALSO gotten more extreme weather conditions.

Obviously if you have global WARMING, you would expect more extreme hot weather.

BUT you also get more extreme COLD weather as well due to the increased variation from the greater energy input as a result of more thermal energy being held in the earth's atmosphere because of the greenhouse effect of the additional CO_2.

People who claim that this is a contradiction of global warming are basically just revealing their lack of education on not only greenhouse gases and how they work but also basic statistics as well.

9. *"Scientist are full of crap. Back in the 1970's they were saying a new Ice Age was coming. That turned out to be wrong. Now they are claiming the earth is warming. You can't believe them"*

Political commentator (read non-scientists) George Will has often said this on TV.

Actually scientist have known about the possibility of not only global warming but mankind's contribution to it going back to the late 1800s

We have already mentioned Swedish chemist Svante Arrhenius and his law that relates CO2 to air temperature. That was in 1896.

In 1959, a majority of climatologist were of the opinion that global warming was a real possibility and that man was probably a big contributor to it.

But it was in 1995, that the Intergovernmental Panel on Climate Change (IPCC), a group of worldwide scientist from over 195 countries, who met after having dedicatedly studied the topic since 1988 concluded as a group that yes indeed global warming was real and that humans were the biggest driver of it.

This group has met many times since then and each time have come away even more convinced of the validity of global warming and man's contribution.

While it is indeed true that a FEW scientist in the 1970s inquired about a new Ice Age possibly developing, they were in a vast minority and they did not hold this conviction deeply or express it as some sort of absolute.

Critics will pick out a few of this scientist from history and hold them up as a "straw man" type of effort to discredit all climatologist of today.

The idea that a majority of environmental scientist thought an Ice Age was soon coming back in the 1970's is just a myth.

In any field, there will usually be a group of people or perhaps even an influential individual who may speculate or predict some event based on known science at the time.

This how does not mean that the entire field is suspect and that any and all predictions must therefore be false not just then but at any time in the future.

That's not how science works.

The fact of the matter is, that there was never a consensus in the 1970s about an Ice Age. There was never any repeatable, reproducible volumes of data like today that would support that assertion,

It's a worthless insignificant criticism.

10. *"Global warming is just a hoax invented by the Chinese to hurt U.S.'s manufacturing base"*

This is from a tweet by president elect Donald Trump although others have stated similar sentiments. (In fairness since the election Trump has walked back this to some extent).

Even as a conspiracy theory this doesn't make much sense.

Putting aside all the aforementioned data that can be independently verified that has already been discussed in this book- how and why would the Chinese do this?

China is the second and soon to be first country in CO_2 emissions. If global warming were real and indeed fully embraced by all countries of the world, then it is THEY who would be most hurt by it. More so than the U.S.

The U.S. over the past two generations has lost a lot of our manufacturing base due to economic reason not environmental ones relating to global warming. (Environmental more often than not is a drop in the bucket of costs compared too much higher items such as work force labor).

China however has built up its economic might via manufacturing. In fact as of today China is basically the world's low cost go-to manufacturer.

This all means that China stands to lose a lot if more regulations or restrictions are made on manufacturing- a big contributor to CO_2 emissions. Picture the big factory smoke stacks bellowing out smoke).

If other countries decided to ban together and not trade with polluters then China is hurt the most.

Why then would China try to spread a hoax that, if fully believed, would in all likelihood come back to hurt them very badly?

This makes no sense at all even by conspiracy theories.

It should be obvious to most readers by now that a lot of this criticisms are politically based not scientifically based.

Ironically the critics of global warming will claim that it is politics that drive the science. They claim that scientist are basically socialist and this is just an end-run type of game to socialize a lot of industries and parts of the economy by claiming some sort of pollution. An attempt, it is implied, to try to socialize private property in a way that the socialist couldn't do in a straight forward manner via the ballot box. (Or worse)

But nothing could be further from the truth.

Climatologist only go where the data takes them. The data from ice core samples. Data which, I will repeat, can be verified by independent objective third parties in a repeatable and reproducible way. This is the hall mark finger print of true science.

WRAPPING IT ALL UP

Hopefully I have explained the science behind global warming theory in such a way that a non-science inclined person could easily follow. It is not important that the average American citizen understands all the minutia details of the science, but it is important that the general theme of where the theory came from and on what basis is it so that they can tell when they are being lied to or not by politicians and other policy makers.

No it's not just based on how hot it is the last few years, although that is consistent with the theory.

No the scientist aren't socialist making a political point by making up pseudo –science to pull the wool over the eyes of all hard working entrepreneurial

No it's not just normal earth variation.

It's based on hard physical evidence and science that can be replicated by other scientist from around the world

It's based known scientific principles on the greenhouse effect that goes back over 120 years.

It's based on H20 behavior that is well known and its fingerprint in ice core samples that go back 800,000 years.

It's based on known CO2 isotopic behavior that is well known.

Although it is not a way to settle the issue, it is no coincidence that nearly 99% of all peer reviewed and published scientific papers over the past 20 years have been involved in confirming global warming theory

Recall that a theory can never be proven true outright, global warming has been but through a lot of criticisms and tests over the years and as of today it is by far the best explanation of what we are observing and measuring in today's earth's atmosphere.

Want to defeat it—come up with evidence, facts, methods, measurements that are BETTER.

THE END